# CHAPTER 1

## INTRODUCTION

Automobile, hybrid technology, innovation, strategy IntroductionHEVs are seen by some researchers as a very promising near-term technology for improving fuel economy and reducing emissions. Proponents also argue that HEVs can provide improved performance for the customer and, in contrast to other advanced-technology vehicles, require no extensive new infrastructure. With many of the advantages but without the range limitation of electric vehicles, HEVs could have broad customer appeal. HEVs, however, come in many different configurations, and even HEV proponents disagree among themselves, which of these is "best." This question motivated the Hybrid Electric Vehicle Working Group (WG), a cooperative effort of HEV stakeholders, to study the prospective efficiencies, emissions, costs, and customer acceptance of different types of HEVs, for a systematic comparison of HEVs with each other and with a conventional vehicle (CV) of similar design and performance. This report summarizes the study approach and key findings of the WG. The study grew out of the discussions of an informal working group that in 1999 brought together knowledgeable individuals from the utility and automotive industries, regulatory agencies and consultants, and the Department of Energy. The initial working group set itself the objectives to establish what was known about hybrid vehicle characteristics and impacts based on credible sources of information; identify gaps in existing information, and define the research needed to fully characterize the different types of HEVs and compare their prospective benefits and impacts. The working group also decided to serve as an initial forum for discussion of the rather diverse views and interests of the stakeholder members in the HEV area. It was envisioned that this group would be able to identify possible strategies and alliances for development, commercialization, and infrastructure support of hybrid vehicle propulsion systems and vehicle options. Finally, the expectation was that the work of the group would lead to increased public and private understanding and, if appropriate, support of all aspects of hybrid electric vehicle system development. The first output of the WG's activities was an informal report, produced with the assistance of ARCADIS (now part of Arthur D. Little), titled Assessment of Current Knowledge of Hybrid Vehicle Characteristics and Impacts, that summarized the results of the survey of existing studies. Although valuable in collecting data and other information on HEVs, the information proved inadequate for a systematic comparison of different types of HEVs. In particular, it was left unclear how the efficiencies and emissions of HEVs deriving part of their propulsion energy from electricity supplies compare to those of HEVs that do not plug in; whether consumers would save sufficient operating cost from plugging in to pay for additional battery cost; and whether customers would see the plug-in feature as a disadvantage or as an advantage by eliminating many or most trips to gasoline stations. Each member had different interests in wanting to learn more about the different types of HEVs.

From the survey results and deliberations of the WG, a conceptual framework (see Figure1-1) emerged for the needed systematic analysis of HEV architectures, quantification of their environmental and efficiency advantages, estimation of their likely future costs, and assessment of HEV prospects for widespread acceptance by customers. To implement the study framework, the WG defined and developed specific work statements for the following four tasks:

1. Modeling of representative HEV types, to ascertain the vehicles' potential for competitive performance, and to determine their emissions and efficiency characteristics for the vehicles themselves as well as for their fuel/energy supply infrastructures over driving patterns/cycles of primary interest.

2. Estimation of key HEV component and vehicle costs and life cycle costs for comparison of HEVs with each other and with a baseline conventional internal combustion engine (ICE) vehicle.

3. Assessment of prospects for customer acceptance of HEVs by prospective owners and users, based upon assumptions about the vehicles' performance and other key driving characteristics, costs, and infrastructure availability.

4. Identification and analysis of likely commercialization issues for the introduction and broad acceptance of potentially beneficial and user-acceptable HEV types, and identification of policy incentives and strategies to mitigate these issues. It should be noted, however, that the WG did not intend to, and has not, taken any position on the merits of particular government regulations, programs or policies related to HEVs

Hybrid electric technology has become the latest milestone forthe automotive industry such have been diesel technology and the gear system in the past. The growing threat of global warming, excessivepetrol dependence, everincreases prices in fuel, and driving trends are just a selection of reasons whichhave accelerated thedeve10pment of Hybrid Electric Vehic1es (HEV). AIso, some government backing has offered supportto HEV technology with the introduction of restrictivelegislationparticularlyconcemed with thereduction ofCOz emissions.

Hybridization of the automotive drivetrain attempts to combine the low emissions of electric automobiles with the extended range of gasoline engines. A hybrid electric vehicle (HEV) increases the fuel economy and decreases the emissions of the system when compared to a vehicle functioning only on a gasoline engine. The greatest benefit of the gasoline engine is the high energy density2 of gasoline, on the order of 12,000 Wh/kg, in contrast with the much lower energy density of batteries, on the order of 500 Wh/kg. This allows the much greater range of vehicles run on gasoline engines. The benefits of electric motors include high torque at low speeds, the absence of on-board emissions, and regenerative braking. Traditionally, there are two ways to configure the system, series or parallel

**Series Configuration**
In a series configuration, the gasoline engine is connected via a generator to the electric motor, and only the electric motor provides power to the wheels. Torque produced by the gasoline engine generates electric energy in the generator, which is stored in the battery for use by the

motor. In this system, the gasoline engine often runs continually in its zone of highest efficiency or lowest emissions, eliminating transient operation of the engine. Numerous types of control strategies are being employed with series configuration. The gasoline engine can be controlled to optimize either fuel consumption or emissions production.

Design of the generator-motor system takes into consideration whether or not the car will be "charge-dependent" or "self-sustaining." A charge-dependent car relies on external electricinput whereas a self-sustaining car does not. The charge-dependent car, thus very similar to a pure electric vehicle, releases fewer emissions; but the self-sustaining car demonstrates a longer running range. Of the two, the self-sustaining car requires a generator of a larger capacity and the charge-dependent car requires a battery of a larger capacity. There are a number of other factors to be taken into consideration in the design and control of series hybrid electric vehicles. The engine does not have to run consistently throughout a driving cycle; thus, the number of times that an engine is started over the cycle is an important variable in influencing the production of emissions.

Another factor is the relation of the battery's state-of-charge and the traction motor output to the input from the gasoline engine.

**Parallel Configuration**

In a parallel configuration, either the gasoline engine or the electric motor, or both can supply torque directly to the wheels. As a general principle, the electric motor is used for starting and low vehicle speeds, and the gasoline engine provides the power for steady-state operation. This configuration presents the designer with an even greater number of design options than the series configuration. Control and control strategy are thus very important.

Control systems function primarily to match the drivetrain with the driving conditions. Some principles are common to most parallel control systems. For example, the gasolineengine is never allowed to idle. When the vehicle is stopped or when it is decelerating, theengine is shut off. Only the electric motor provides torque for all slow-moving operations. A minimum vehicle speed is usually set to govern the entrance of the gasoline engine. Both the gasoline engine and the electric motor are used together for operations that demand hightorque. Regenerative braking is employed.

A number of factors vary among designs. Designers must choose a minimum speed below which the gasoline engine is turned off. They also determine a minimum operating torque as a function of engine speed for the gasoline engine . If the torque required to meet the trace, which is the instantaneous torque demand on the vehicle, falls beneath this mark, the excess torque is used to drive the motor as a generator, recharging the batteries. A parallel-configuredhybrid can run the gasoline engine in a number of ways; the gasoline engine can be used to meet the trace, it can be used only for steady-state operation, or there can be an intermediate control strategy.

Comparison of the Two Control Systems

Control strategy in both series and parallel configuration is a significant determining factor for the operations and performance of hybrid electric vehicles. Variations in strategy can produce large variations in emissions production and fuel economy.

The parallel configuration is being most commonly chosen by automobile manufacturers. The operation of parallel hybrids more closely resembles the operation of traditional cars than does the operation of series hybrids, thus rendering them more appealing to the consumer. Also, the parallel hybrid has been shown to be 4% more fuel efficient than the series hybrid, primarily because the gasoline engine supplies power directly to the wheels, converting from mechanical power to electrical power and back again, as occurs in the series hybrid.

# CHAPTER 2
## HEV HISTORY

The competition between vehicles powered by electric and those powered by an internal combustion engine (ICE) is notanew scenario; this antagonism dates back to as early as the beginning ofthe 19th century. Between 1890 and 1905 ICEs, electric vehicles (EV s), and steam powered cars 5 were all marketed in the United Kingdom and United States. EV s were the market leader in the United States atthis time; mainly due to the works of electricity pioneers such as Edison and Tesla. The limiting range of EVswasnotabigproblemas the roads linking the cities were not particularly adequate forvehicle transportation.

It was evident that the use ofbatteries in automobiles was going to pose limitations inrange and utility ofEV s. Due to the energy advantages of petrol powered vehicles over battery operation, petrol became the dominate energy source overthe next 100 years, andis stillleading the waytoday. At the time many automotive companies designed direct ICE vehicles, but sorne tried to combine the advantages ofthe electric vehicle with those of an ICE vehicle by creating a hybrid ofthe two.

The first ever REV was built in 1898, and therewere several automotive companies who were selling REV s in the early 1900s. The production of HEV s didnotlastthecourse oftimedueto significant problems with them. H enry F ordinitiated themass production of combustion enginevehicles; making them widely available and affordable within the $455 to $911 price range (H» 375€ to 750€ with prices taken from the current American dollar to

Euro conversion rate). In contrast, the price ofthe less efficientEV s continued to rise. During 1912, an electric roadster sold for $1,732 (1 ,425€ ), whilsta gasoline car sold for $547 (450€ ) as illustrated by About Inventors. Another problem was the requirementfor a smooth coordination between the engine and the motor, which was not possib le due to the use of only mechanical controls.

Since these early attempts, there has been a rise in the concern for global warming, a continual rise in fuel prices, and the threat of oil reserves dryingup altogether. This hadled to interest in more efficient and environrnentally means oftransport again,particularlyin theareaofHEV. Withadvances in battery technologies and onboard computer systems, the option of a plausible HEV has become reality, and a number of models from the likes of Honda (Civic and Insight) and Toyota (Prius) have been available now since 2000. There have been a number of prospective designs andREV shave beengrowing eversince the inclusion ofthem onto the world market in 2000. The increased interest along with legislative movements has made advanced clean and efficient transportation notonly a vision forihe future, but one fortoday

This article deals with the hypothesis that the recent growing craze for hybrid vehicles in the United States and Europe is simply a temporary step between the traditional technology based on gasoline and diesel engines and the forthcoming of full electric vehicles probably with hydrogen powered fuel cells. Such assumption is shared by several observers from professional as well as academic background (Ashley, 2002; Hekkert, 2004). Hekkert (2004) is the most radical challenging the idea that the emergence of hybrid vehicles might be at the expense of the fuel cell vehicle. Chanaron and Orselli (2002) suggest that hydrogen fuel cells will not be marketable in high volumes before at least 2025 and that most if not all information released so far are pure manipulation and marketing by the hydrogen lobby. The quest for low emission (clean) and high mileage vehicles is on its way and will surely remain at the top of the OEMs agenda.

Because new facts and events occur on a daily basis, such an article is inevitably out-dated as far as factual information and data are concerned. They have been up-dated up to the end of 2006. It has to be pointed out that the research is targeting only passenger cars, SUV and light commercial vehicles.

The present article is based on the following set of statements that it aims at illustrating and discussing:

- There is a general convergence of strategies towards promoting hybrid vehicles as the mid-term solution to very low emission and high mileage vehicles;

- Such a convergence is largely due to Toyota's strategy learning the technology while building up its own "quasi-standard", thanks to its high quality and reliability reputation and its high market share on the North American market;

- Such a convergence is based more on customer perception triggered by very clever marketing and communication campaigns than on pure rationale scientific arguments and may result in the need for any OEM operating in the US to have a HEV in its model range in order to survive.

Obviously, such statements lead to several unresolved questions: Is such a strategy sustainable on the long run? Or is it a short to mid term option which would then last only a few years and thus remain to limited manufacturing volumes? What are the triggers to technical choice? Is it the lowest $CO_2$ emission or a more complex efficiency mix between cost, pollution and energy?

The economic and managerial literature dealing with hybrid vehicles is still very poor, basically limited to enthusiastic press coverage for newly launched models or surveys on cost efficiency (Chanaron, Faudry, 2005). Most academic publications emphasize scientific and technical issues, such as power train modeling and control, braking systems with regenerative devices, energy storage, energy management systems, propulsion systems, electric engines, vehicle simulation, vehicle design, etc. Most key articles limit their scope to technical efficiency (Demirdöven & Deutch, 2004).

Within the available literature, as surveyed by Chanaron & Faudry (2005), most publications deal with general discussion about the potential technologies and innovations limited to qualitative analysis of key drivers and success or failure factors. Most information comes from professional materials, press cuttings and interviews of key researchers and decision makers and also from previous research (Chanaron & Nicolon, 1976; Nicolon, 1977; Chanaron, 1983, 1994, 1998). They highly reflect the point of view of industrial actors of the automotive system (de Banville, Chanaron, 1991), not to say the automotive lobby. There is hardly any interest given to consumer acceptability, pricing, etc. Amongst these qualitative surveys, there are also historical essays such as Bardou and alii (1982) and general reflections on the future of the automobile technology. The vast majority of articles and books belong to industrial economics. Very few refer to management sciences.

Struben & Sterman (2006) start from the historic failure of the electric car at the end of the 19th century to built a model to explore the so-called transition between two technologies, i.e. internal combustion engine (ICE) and alternative fuel vehicle (AFV), including hybrids, natural gas and hydrogen fuel cell vehicle. The authors introduce the concept of familiarity in order to capture the "cognitive and emotional processes through which drivers gain enough information about, understanding of, and emotional attachment to a technology". Awareness and familiarity are correlated with the rate of adoption. Familiarity and consumer choice is highly dependent on marketing, direct social exposure, indirect word of mouth. They conclude that the transition will be longer than usually expected and will involve a wide array of interactions and feedbacks which are not yet captured by the existing models, such as interactions with other industries and the fuel supply chain.

As far as market-related studies are concerned, it is worth mentioning Kishi & Satoh (2005) and Sanchez-Repila & Poxon (2006).

Kishi & Satoh (2005) explore the evaluation of willingness to buy a low-pollution car in Japan based on a price sensitivity measurement model through a questionnaire survey run in Tokyo and Sapporo. Their research shows that the correlation between environmental awareness and willingness to buy is high as well as not surprisingly the relationship between decrease in price and willingness to buy. Their most interesting result is their evaluation of the minimum, maximum, standard and reasonable price for hybrid cars.

Sanchez-Repila & Poxon (2006) present the different technical alternatives but also devote a specific section to market status, examining the specificity in USA, Japan and Europe. They point out the necessity of tax incentives in order to compensate for the higher prices. They also emphasize the fact that the customer motivation to buy a hybrid car seems to be environmental rather than economic, the savings in fuel being not so significant. Hybrids seem to benefit from a strong environmental and technological image or perception. The authors conclude with the obstacles to future developments of hybrids, and in particular the cost of shifting to 42V supply, and they foresee no volume take off before 2010.

## CHAPTER 3
**The Hybrid Vehicle and hybrid system**

Any vehicle that combines two or more sources of energy that can provide propulsion power, directly or indirectly, is a hybrid. They are usually vehicles that use an internal combustion engine and an electric motor as an alternative source of energy. For this study will be considered a hybrid vehicle, a light vehicle with an electric motor and an internal combustion engine powered by gasoline. Hybrid vehicles run on an internal combustion engine, but are also able to convert energy into electricity that is stored in a battery until the electric motor is in operation, pulling the vehicle, thus saving the energy required by the internal combustion engine. This allows the internal combustion engine is more efficient, use less fuel and thus produces fewer pollutants. Therefore, the electric motor is used when the internal combustion engine efficiency is low, i.e. when accelerating when going up a hill, when at low speed, pulled or when the car stops. In breaking the gasoline engine automatically shuts down and the electric is turned on, and through regenerative braking the battery can be recharged. Thus, the electric motor kicks in when it is to save fuel. Unlike electric vehicles, hybrid vehicles do not need to be connected to external sources of electricity, using only energy from the combustion engine and regenerative brakes.

The combination of two sources of energy is more efficient than the internal combustion engine or electric motor alone. Hybrid vehicles can be configured in several ways, combining the internal combustion engine is better with the electric motor assist, improving fuel economy and reducing pollution without sacrificing ride quality and performance.

**Hybrid systems**

As already described in this section, hybrid systems use a combination of internal combustion engines and electric motors, and are made in its basic structure by a gasoline engine, fuel tank, electric motor, generator, batteries, and transmission. A brief characterization of each one follows.

The hybrid car has a gasoline engine very similar to that found in most cars. However, the engine of a hybrid is smaller and uses advanced technologies to reduce emissions and increase efficiency. The fuel tank in a hybrid vehicle is the storage device to power the gasoline engine. Gasoline has an energy density much higher than that of batteries. For example, it needs 450 grams of batteries to store the same energy generated by 3.79 litres or 3 kilograms of gasoline.

Electric motor in a hybrid car is very sophisticated machinery. The advanced electronic technology to its role as both a motor and as a generator. For example, if it needs this kind of car can draw power from the batteries to accelerate. However, acting as a generator, it can stop the vehicle and return energy to the batteries. Generator is similar to an electric motor, but acts only to produce electricity. It is mainly used in hybrid as series arrangement. The batteries of a hybrid car are the storage device to power the electric motor. Unlike gasoline in the fuel tank, which can trigger only the gasoline engine, the electric motor in a hybrid car can provide power for the batteries and still draw energy from them. Transmission in a hybrid car performs the same basic function that the transmission in a conventional car. Some hybrids, like the Honda Insight, have conventional transmissions. Others, like the Toyota Prius, have very different transmission, which uses a system of division of power, consisting of planetary gears.

Is possible to combine the two sources of energy found in a hybrid car in different ways. The configurations of hybrid vehicles can be three different types: series system, parallel system, and combined system. In the series system configuration the internal combustion engine is used as a generator providing power to the electric motor and battery. The internal combustion engine is not mechanically coupled to the wheels, and may be controlled, thus generating efficiency optimisation control of reaching emission levels, regardless of their driving conditions. In this case, the gasoline engine does not move directly from the vehicle, only the electric motor drive the wheels. In the parallel system configuration the internal combustion engine is mechanically coupled to the wheels being able to supply the required power. The electric motor is mounted parallel to the internal combustion engine so that it can add the torque required for its operation. The internal combustion engine can then drive the electric motor as a generator, thus charging the battery. This system does not require a generator as the series system. Both the gasoline engine and electric motor can turn the transmission at the same time and the transmission then turns the wheels. In the case of series-parallel combined system this one uses the characteristics of systems in series and parallel together; requires both functions, a generator and a motor.

# Chapter 4

## Working principle of HEV

Hybrid electric vehicles attempt to capitalize on the complementary characteristics of the internal combustion engine and the electric motor in order to minimize fuel consumption and the production of emissions. There are numerous ways to configure the system and strategize the control system. The following is a generic control strategy. It is chosen both for its characteristics of minimal fuel The following is a generic control strategy. It is chosen both for its characteristics of minimal fuel consumption and emissions production, and because this system can make the HEV perform in a very similar manner to the traditional car, an important reason for widespread acceptability in the

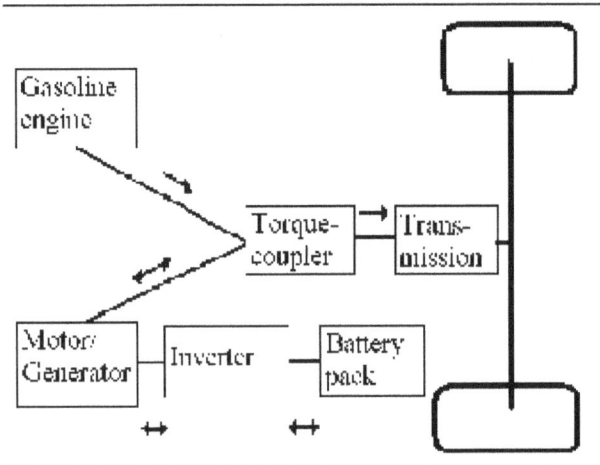

Schematic diagram of a generic parallel control

Figure shows the schematic diagram for this control strategy. In this parallel-configured system, both the engineand the motor provide torque to the wheels. For the most part, the internal combustion engine isused for the high-energy demands on the car, because of the relatively high specific energy levelof gasoline. The engine is set to operate as much as possible in the region of lowest brakespecific fuel consumption, and to minimize transient engine operation. The electric motor is used mostly for the high power demands on the car, because it produces high torque at low speeds, and because it demonstrates a high drivetrain efficiency over the its range of torquesand speeds. When the motor is not in use providing motive power, it can be run as a generatorfor two purposes. In regenerative braking, the generator converts torque from the wheels inorder to decelerate the vehicle and store this energy in the battery. The vehicles are also equipped with friction brakes for safety purposes. Also, the generator can convert torque from the gasoline engine to electrical current; this is done to maintain the state of charge of the battery to a desirable level, and also, to demand higher torque from the engine to keep it in its zone of highest efficiency. A system of electronic controls (equipped with computer technology) and mechanical controls matches the driving conditions for the vehicle to the appropriate drivetrain or combination of drivetrains.

The following represents a typical system: 1) starting the vehicle, (speeds from 0 to approximately 10 mph): electric motor only (in this way, the engine is restricted from idling and low speeds, part-load conditions where combustion efficiency is very low); 2) braking: engine is shut off and motor runs as a generator (engine is again restricted from idling, and energy is recaptured and stored that would be wasted in a vehicle without regenerative braking capacity); 3) cruising conditions: engine only; 4) rapid acceleration and climbing hills: engine and motor together to account for elevated power demands. In this fashion, hybrid electric vehicles have been shown to double fuel economy over a comparable traditional automobile and halve the production of emissions.

No universal control strategy is in use. The design of HEVs permits great flexibility, allowing the designers to optimize for a number of different benefits, such as fuel economy, emissions, cost of the vehicle, and safety. Comparison of the three HEV concept cars produced by the automobile manufacturers and displayed at the January, 2000 Detroit Auto Show reveals this large variability in design. In the GM Precept, the electric motor powers the front wheels and the gasoline engine powers the rear wheels. Toyota employs the "Prius Hybrid System" using a continuously variable transmission, and both an electric motor and a separate generator. Hybrids could be programmed to learn the driving patterns of its owner and adjust to them for maximum improvements in emissions and fuel economy[11]. A number of other systems are possible.

# CHAPTER 5
## Peoples intention on buying hybrid cars

Consumers are buying increasing numbers of environmentally friendly cars. Increasingly, many of these environmentally conscious consumers choose to purchase petrol-electric hybrid vehicles. In this category of "greener-cars", Toyota's Prius model is reported to be the market leader. In 2009-10, it was the best-selling car in Japan, an important leading market for automobile trends (Mick 2010). Sales of the Prius keep growing despite well-publicised quality and safety problems (Mitchell & Linebaugh 2010). In fact, the demand for petrol-electric hybrids is so strong that Toyota has introduced a second and larger Camry branded hybrid vehicle into Australia. Other car manufacturers are following with their own models, indicating that there is likely to be sustained demand for this type of light-duty passenger vehicle.

Toyota markets the Prius as an environmentally better alternative to conventional vehicles because it uses less fuel and has lower emissions. This marketing position appears to appeal to consumers who do not wish to further degrade the environment. It has been suggested that these consumers choose to help by driving a car that is more environmentally friendly (Griskevicius, Tybur & Van den Bergh 2010; Bamberg 2003). Popular sentiment has it that intrinsic motives to preserve the environment are the driving force behind the popularity of these vehicles. This is because consumers keep buying petrol-electric hybrid cars like the Prius even though they cost more then twice the amount of a comparable conventional car. But are intrinsic reasons really why consumers choose to buy a car like the Prius? Are there other reasons behind its popularity?

It has been recognized that encouraging the adoption of environmentally friendly products is a key challenge for the behavioural scientists (Kaplan 2000). This appears to be why there has been a great deal of research into the reasons behind this adoption. This article seeks to add to this knowledge by exploring the reasons that drive adoption of environmentally friendlier automobiles, specifically, the petrol-electric hybrids that are gaining popularity. This information may potentially be valuable to increase adoption rates for other environmentally friendly products and ideas.

## Environmental sensitivity and consumption

The currently popular paradigm for discussing the environment originated in the 1970s, when the ideas of global warming and finite oil reserves were first proposed (Minton & Rose 1997; Pelletier et al. 1998). While some debate continues on the veracity of these propositions, this thinking has influenced the way people live by increasing their efforts to reduce energy use and to have fewer by-products as a result of consumption. It has been suggested that this type of thinking has led some consumers to prefer products like the Prius (Jansson, Marrell & Nordlund 2009). These consumers with ecological and environmental concerns have been described in various ways, and are sometimes called environmentally-sensitive, -conscious, or as environmentalists. This group of consumers are reportedly more positively oriented towards conservation and environmental issues when compared to other consumer groups (Casey & Scott 2006; Minton & Rose 1997; Stern et al. 1995) and are widely documented as having a higher tendency to adopt eco-friendlier products (Gatersleben, Steg & Vlek 2002; Minton & Rose 1997; Anable 2005; Bamberg 2003; Hansla et al. 2008; Maloney & Ward 1973; Stisser 1994). This hypothesis has also been tested in its inverse, leading to the finding that consumers who were inclined towards eco-friendlier products were also the most sensitive to the environment (Jansson, Marell & Nordlund 2009). The literature into reasons for buying environmentally friendly products appears to be split along the streams of intrinsic versus extrinsic motives. The methodology employed appears to be correlated with the finding for either intrinsic (unmasked data collection methods) or extrinsic (masked/disguised data collection) motivations for adoption.

**Intrinsic motives**

Intrinsic motivations appear to be the reason why conservation and environmentally minded consumers adopt eco-friendlier products (Chan 1996; Bamberg 2003). For example, intrinsically driven consumers buy hybrid cars to reduce the effects of their driving on the environment. However, there is another stream of literature using other research methods (often masked surveys) that have found that pro-environmental products are often not purchased because of pro-environmental motives (Diekmann & Preisendörfer 1998, Barr 2004, Mainieri et. al. 1997). In fact, in a wide-ranging study that employed disguised surveys on a range of products, Bamberg (2003) reported only a low to moderate association between consumers' concern about the environment and adoption of consumption behaviour that was considered to be environmentally friendly. If this is the case, is it possible that there are other reasons behind adopting 'greener' products?

**Extrinsic motives**

It is also possible that extrinsic rewards (e.g. popularity, image, status) may be a more salient reason for some consumers to adopt environmentally friendly products (e.g. Jansson, Marrell & Nordlund 2009; Stern 2000, Clark, Kotchen & Moore 2003). This is not to say that these consumers do not posses intrinsic motivations, but that extrinsic reasons appear to play a more powerful role in their decision making process.

Griskevicius, Tybur & Van den Bergh (2010) conducted a series of experiments and found that many of the consumers' in their sample chose environmental friendly products because of social or professional status concerns. In addition, participants were more likely to adopt environmentally friendly products with 'conspicuous' consumption characteristics (i.e. they could be seen using the product) over those that were predominantly consumed in private. They also found that consumers were more likely to choose environmentally friendly products that were comparatively more expensive, and tended to shun those that were priced at the same level or below similar less eco-friendly products. Griskevicius et al. found positive correlations between price and public consumption, suggesting that the consumers tested in their experiments appeared to view adopting green products with social and professional status. Additionally, they only adopted green products that were more expensive (i.e. luxury green products). The authors indicated that their results clearly suggested that some consumers tended to adopt an environmental friendly product only if this consumption decision was advantageous to their image as being pro-social issue and unselfish. While Griskevicius et al.'s study investigated the influencing role of status in the adoption of eco-friendly products, no attempt was made to test for the relative the importance the different factors that enter into the decision making criteria for products. Nor did they compare the decision making criteria for the purchase for non-environmental products against those used to evaluate environmentally friendly products. Our research adds to the literature in this area.

In order to maximise the differences that can be discussed, most academic research dealing with contrasts prefers to use groups with strong tendencies towards the end-points of the phenomenon (e.g. highly intrinsic vs. highly extrinsic groups), there is also likely to be a third group of consumers with a more equal mix of in- and extrinsic motivations. However, because of the often non-significant result that is obtained from this group with more balanced views, these results are normally not reported in academic literature (c.f. Clark, Kotchen & Moore 2003). It is likely that this is the largest group of consumers when it comes to adopting hybrid products.

**Related research in choice of vehicle fuel**
An area that is related to a consumer's choice of car is the choice of fuel. Four thousand Swedish drivers were surveyed on their level of eco-sensitivity and the type of vehicle fuel they used (Jansson, Marell & Norlund 2009). As expected, most were not sensitive to environmental issues and did not adopt less polluting fuels. What is surprising is that consumers who reported the most sensitivity to environmental issues also reported driving the least, instead preferring public transportation. Unfortunately, Jansson et al.'s study did not report if this group was also more likely to adopt "green-cars". Neither did Jansson's study measure the in- or extrinsic motivating factors behind adoption. While the study reported a correlation between adoption of cleaner-fuels and level of eco-sensitivity, the results are only descriptive in nature, offering little insight into why people choose more ecologically sound products. Our study tests if motivations (in- or extrinsic) play a role in the way consumers select cars for purchase and attempts to identify the type of motivations that operate in the choice of (non) environmentally friendly cars. At this juncture of the discussion of the extant literature, it is appropriate to cover the way consumers make decisions.

**Consumer decision making process**

For many consumers, choosing an automobile is often a complicated and high-involvement process. Although cars are regularly used products, they are also rarely bought products. Additionally, an automobile is expensive, there is a large selection and the consequences of not choosing well typically lasts a long period of time and may cost a lot to rectify.

Consumers enter into the process of actively evaluating automobiles for purchase when they experience a strong desire or need for a car (Dholakia 2001; Frey & Jegen 2001; Villacorta, Koestner & Lekes 2003). Coupled with the ability and desire to buy, the consumer is said to be "in the market" for a new car. This means that the consumer is saving money or has access to funds for purchase, and they have strong intentions to complete the purchase in the near future. We have adopted this definition of 'in the market' for purchase to select the sample for our research.

To help them arrive at a final choice, many consumers will weigh and evaluate different factors. While the factors that are evaluated and their importance are expected to differ for individual consumers, as a group, they are expected to take into consideration some common elements. These can include cost, practicality/performance, aesthetics, the 'lifestyle/image' associated with some makes/models, social influence and the car's environmental credentials like fuel economy/emissions (Griskevicius et al. 2010).

Consumers use weighting of these factors under consideration as a form of shorthand to make their decision making process easier. They use the degree of importance of the vehicles' elements or groups of elements that form factors to help them reduce the items that are being considered. These weights can be seen to form key consideration sets and are likely to finally be used to establish "purchase parameters". These purchase parameters tend to form the final decision-making or –breaking standards for purchase.

In the evaluation process, the consumer seeks vehicles that best address the decision-parameters that they have established. A popular method for portraying this process is as a trade-off between different factors and dimensions.

If this trade-off process is driven by underlying intrinsic or extrinsic reasons, then it is likely that these reasons will manifest themselves in the factors or dimensions that are used by consumers to evaluate vehicles for purchase. These dimensions are sometimes called consumer consideration-sets. In this sense, a person who is buying a car to "show-off" may consider very use very different consideration-sets to those used by a consumer who is intrinsically motivated and buys a hybrid vehicle because they wish to save the environment. This kind of decision making mechanism affords researchers the opportunity to study the factors that enter into consideration for the purchase of cars. It may even be possible to compare the choice sets used for the purchase of different models or types of cars (e.g. conventional vehicles vs. petrol-electric hybrid cars).

At this point, it is useful to introduce the possible dimensions and constructs that consumers may use for choosing between different automobiles. Product performance/function includes evaluations of how the product is likely to perform (Lavidge & Steiner 1961). Utility is a common measure product performance. The consumer can evaluate performance first hand by test-driving the car or may obtain it second-hand through the media or through word-of-mouth. Product grade is closely related to performance and is the product's perceived quality and attributes (Lavidge & Steiner 1961; O'Brien 1971). Buyers may believe that hybrids produce lower emissions, making it a better quality automobile. Typically, product quality is negatively associated with product price.

The cost of a car includes its purchase price and running costs. In Australia, a Prius costs $16,000 more than a Corolla, which is a comparable car (Toyota Australia 2010). Running costs for hybrids are also generally higher. Is this price premium a consideration for hybrid buyers? The cost of a car has been found to be positively associated with its perceived image (Heffner, Kurani & Turrentine 2005, 2007). For example, expensive cars are perceived to be prestigious and luxurious. Indirectly, this prestige is transferred to its driver. Similarly, Prius drivers may derive benefits from the "green" image associated with hybrid vehicles.

Social influence can affect an individual's choices (Ajzen 1991). People use specific products in order to gain admittance, fit in with, and to attain social standing within desired reference groups (Steg 2005; Heffner, Kurani & Turrentine 2007; Pelletier et al. 1998; Griskevicius, Tybur & Van den Bergh 2010). This factor has been found to be significantly stronger for some groups of consumers (Steg 2005). Typically these consumers are described as conformists who will follow the directions and wishes of their referent groups. All of these dimensions help the consumer choose which car to buy.

## Limitations

This article has reported the results of an investigation into the latent dimensions used to evaluate car purchase by consumers who were considering buying a hybrid-electric car. These dimensions were compared to a group of respondents who were thinking of buying a conventionally fuelled vehicle. These two groups of car buyers were actively evaluating automobiles for purchase in the next twelve months and had saved money or had access to funds to purchase a vehicle.

We found that respondents choosing hybrid vehicles evaluated the purchase differently from buyers choosing conventional vehicles. At least in this sample, hybrid buyers were mainly concerned with whether the car would improve their social standing and personal image. This finding is consistent with that reported by Griskevicius, Tybur & Van den Bergh (2010). Buyers of conventional cars were more concerned with the car's functionality, cost and quality and were less concerned wither the car made them socially popular. What this article added to the literature is that we now have an idea of the relative importance of the evaluative criteria for petrol-electric hybrid vehicle purchase. Our model also provides a map of how the items have loaded into these evaluative dimensions.

Although the sample consisted of many university graduates, the sampling was done in the workplace. It just happened that many of the researchers' colleagues were graduates. This population was young (22-30) and were the prime market for automobile manufacturers.

It appears that hybrid cars, at least for our sample, appear to be purchased for social and reasons and not by people who genuinely care for the environment. This finding is reasonable as other research (e.g. Heffner, Kurani & Turrentine 2007) have reported that the most environmentally sensitive consumers preferred to abstain from driving. Our sample of hybrid car buyers is highly influenced by their reference groups. These groups seem to dictate the consumption behaviour of Prius buyers.

This strong influence of groups can be utilized by social marketers to shift social behaviour increase adoption of more environmentally friendly cars. In this case, a suitable model appears to be the diffusion of innovations model. Although rarely used by social marketers (Lefebvre 2000), this model can be used to promote the orderly adoption of hybrid vehicles. The way hybrid automobiles are bought, driven by social influence, provides positive answers to four areas necessary for successful diffusion of an innovation (Oldenburg, Hardcastle & Kok 1997). These areas are: does it fit into the audiences' lifestyle and self-image? Is the new behaviour better than current behaviour? Can it be trialled before commitment? Can it be easily and clearly understood? Finally, can the behaviour be adopted with minimal risk?

This externally motivated group (perhaps early adopters) can be influenced to achieve a critical mass of adoption for low-emissions vehicles. Social marketers can do this by using reference group appeals. In this case, it is important to achieve a salient positioning for the concept of low-emissions vehicles so that it appeals to more than one agency, and to later adopter groups that may be less prone to social influence (see results for buyers of conventional cars). It is expected that in order to achieve widespread adoption, the agencies that must be influenced include the customer, policy makers (e.g. government and regulators) and the community at large. It is only with this acceptance that the utilisation of low emissions vehicles will reach a critical point that it provides a positive effect on the environment.

The diverse degree of diffusion of the technology makes it difficult to draw one common global picture on the HEV market. Therefore, an insight to key markets is provided. Sales and registration data exist for the United States as well as for Japan. In the US, HEV sales have risen consistently since 1999:

# CHAPTER 5

## 5 Most Popular Electric Cars

### Advantages

- The number one advantage of an electric vehicle is that no gas is required. One example is the **Chevy Volt**. It has a battery range of 40 miles. That means it can drive for 40 miles without using gas. 40 miles is more than the range of an average commute to work, so you can go to and from work using no gas. With minimal gas usage comes great savings. You do need gas in the Volt in case your battery runs out or you go for a long distance. However, the amount of fill ups per year will be much fewer with an electric vehicle

- You can plug the car into any outlet of the proper voltage and charge the car. Electricity is much cheaper than gas, and the savings will be dramatic

- Electric cars give off no emissions. Electric cars are even better than hybrids in this regard. Hybrids running on gas give off emissions, while electric cars are totally 100 percent free of pollutants

- Safety is a big concern with these vehicles. However, the fluid batteries actually take impact better than a fully made gas car, and can help even more in the event of an accident

### Disadvantages

- The first disadvantage is price. Electric car batteries are not cheap, and the better the battery, the more you will pay. For example, the Chevy Volt has a 40 mile range and sells for around $30,000. Compare that to the 250 to 300 mile range of cars made by Tesla Motors, which sell for anywhere between $50,000 and $100,000

- Even though it is a quiet ride, silence can be seen as a disadvantage. People like to hear cars when they are coming up behind them or beside them, and you can't hear if an electric car is near you. This has been known to lead to accidents

- Most cars take a long time to recharge their batteries. Tesla Motors' Model S can recharge in 45 minutes, but most electric cars right now take hours to charge. You can't drive the car while the batteries are charging usually, so your car will be out of commission while it is plugged in

- Most electric cars currently on the road do not have long ranges. Although in the future it will improve, most of the cars have a range of less than 25 miles, and you can't truly see the great benefits until you ride in a vehicle with a longer range

### Current and Future Electric Cars

Many automakers have plans to produce electric cars--also known as an electric vehicle (EV). Past EVs never gained popularity and had notoriously poor performance, short battery life and long recharge times. But automakers are confident that advances in technology will ensure that the next generation of electric cars will satisfy the needs of today's drivers. Here are a few models scheduled to appear in showrooms soon.

## TeslaMotors

Tesla Motors was founded in 2003 and quickly earned worldwide attention following the unveiling of the Tesla Roadster in 2006. This two-seater sports car accelerates from 0 to 60 mph in 3.9 seconds and has a travel range of more than 200 miles on a fully charged battery. The Roadster's efficiency and performance showcases the company's belief that an electric car can be both environmentally friendly and exciting to drive. The base MSRP is $101,500, which includes a $7,500 federal electric car tax credit. A "Sport" version is also available for an additional $19,500 and includes a sport adjustable suspension, performance tires and a 0-60 time of 3.7 seconds.

For those needing more room, Tesla is currently taking reservations for their upcoming Model S, an electric sedan that the company says will accommodate seven passengers. Varying battery and charging options will give the Model S a range of up to 300 miles and a charge time as fast as 45 minutes. Deliveries of the Model S are scheduled for late 2011 and the estimated price is $49,900 (including the tax credit).

## ChevroletVolt

The Volt is a five-passenger sedan that travels on pure electricity for an estimated 40 miles per battery charge. For longer trips or when charging is not possible, the Volt automatically switches to an onboard range extender, a gasoline-powered generator that creates more electricity to power the car. Chevrolet says the range extender will allow the Volt to travel an additional 300 miles per tank. The Volt is in showrooms at around $40,000.

## FiskerKarma

Fisker Automotive is another start-up that the Karma, a four-passenger luxury sedan that uses a system almost identical to the Chevrolet Volt. The Karma has a 50-mile range on pure electricity before switching to a gasoline-powered generator that provides power for 300 miles per tank. The base price is around $87,000 (before tax credits are applied).

## SmartForTwoElectric

Smart is currently testing and researching an electric version of the ForTwo, which has an estimated range of 80 miles per charge. We should see the electric smart by 2012. Prices haven't been announced.

## NissanLeaf

The Leaf, a five-passenger sedan with an estimated 100-mile range per charge, was released in late 2010. The Leaf sells for around $36,000.

You can expect this list of electric cars to grow as more customers look for alternatives to gasoline due to fluctuating gas prices and concern for the environment.

**Electric vs. Gasoline: The Initial Costs**

Electric cars have decreased in price considerably over the last 10 to 15 years. However, some electric cars still remain very expensive. Today electric cars can be purchased in the $10,000 to $15,000 price range for very basic entry models, and go all the way up to over $110,000 for some of the most advanced models. This is similar to many vehicles that use gasoline engines as well. The similarities and costs seem to stop at the initial purchase price.

**Continued Operation Costs**

With a gasoline-operated vehicle, buying gas in order to drive is a never-ending process. With the ever-increasing cost of gasoline and oil, operating a gasoline vehicle tends to be very expensive. While there are some types of smaller gasoline vehicles that get much better gas mileage than larger SUVs or pickup trucks, even the most efficient gasoline models will generally require a substantial investment every month.

On the other hand, electric vehicles do not require gasoline to operate and use standard electricity plug-ins to charge their batteries. In fact, some electric car manufacturers claim that their vehicles require as little as $0.01-$0.02 a mile to operate. Compare this to an average minimum cost of about $0.08-$0.10 per mile for even the most fuel-efficient gasoline engine vehicles, and you can quickly see how electric cars can offer a lot of savings for daily commutes and short range driving.

**OtherConsiderations**
While electric cars are certainly more efficient and less expensive to operate, and prices have now dropped to levels that rival some many popular gasoline engine vehicles, there are certain things you should consider that may not be able to be gauged in terms of cost or value.

For example, you will not always be able to simply jump into an electric vehicle and take off. This is because electric vehicles use rechargeable batteries that can take considerable time to recharge. In fact, one of the major complaints against the election vehicles is the lengthy charging time that is required. For many electric cars, it can take between 8 to 12 hours to fully recharge an empty battery cell.

Electric vehicles are severely limited in the ranges they can travel. Typical electric cars can only maintain a charge and be driven for distances between 60 and 150 miles. Even the most expensive models tend to top out at a range of about 200 miles. Contrast this with the standard gasoline engine vehicle, and you can see how many people would consider a gasoline-operated car or truck a better choice regardless of fuel costs.

**Electric vs. Gasoline vs. Hybrid Car Pricing**

Of all the information that you are going to want to compile when researching and purchasing a new car, car pricing is probably the most important piece of data you want to have. As one of the most important factors to consider when deciding on which car to purchase, accurate car pricing information is essential to finding the best deal on the car you are interested in. Whether it's figuring out the MSRP of this year's crop of new models or soliciting new car quotes from multiple dealers in your area, having a clear picture of how much your new vehicle is going to cost can help you avoid overpaying and get the car of your dreams at the best price.

Because of the multitude of factors that can go into determining the price of a new car, it helps to have as much information as possible on how these prices are calculated and how you can find a way to pay the lowest price possible. While researching new car prices, take some time to read through the expert articles and advice that we have collected here to help you understand the mechanisms of car pricing and give you strategies for how to make the most of your money.

**New Electric Car Costs vs. New Hybrid Vehicles**

Nearly every manufacturer now offers some sort of version of the hybrid car that combines electric engines with traditional petroleum engines. This allows consumers to use much less gasoline than normal as well as emit a lesser amount of $CO_2$ into the environment. The options for these greener cars have increased over the years to provide not only standard economic cars, but larger SUVs as well. Because of the demand for hybrids, the average cost for each vehicle has actually decreased. Allowing more individuals the opportunity to purchase one, and essentially, saving money in the long term due to reduced trips to the pump.

The hybrid ultimately saves money in fuel and runs for about $30,000-$45,000, depending on the make and model you desire. While they have lower emissions and require less frequent trips to the gas station, you'll still have to pay to keep it running. Maintenance is low, however, gas prices are still increasing in cost. Even though you won't pay as much as the average person, you will still be paying something.

Electric vehicles are actually not really a futuristic concept. Some of the first automobile prototypes only used electricity to run their engines. The only problem with this is that they didn't run very fast, nor did they run for very long without a need for recharging. That's why the traditional engine using petroleum replaced all electronic forms of transportation.

Years later, and with a depleting oil economy, there was a need to revisit the electronic vehicle. To manufacture anything that would be remotely suitable for transit, proved it would cost a fortune for the consumer once completed and ready for market. Once the first marketable electric car came about, it was priced at nearly $110,000. The upside to paying this kind of cash is that there wouldn't be any fuel costs at all for the life of the vehicle.

Now there is not only a reasonable solution to having an electric car, but one that is stylish and sexy that anyone would want this vehicle. Tesla Motor Company created the first, affordable roadster at half the price of previous street friendly vehicles at just under $50,000. Its sleek, its quiet and will never need refueling. It will also travel up to 300 miles without having to need a recharge, which really will only take 45 minutes total to complete. The Tesla Roadster S is the first in its kind to be an electric car and go from 0 to 60 miles per hour in 5.6 seconds, as well as being able to hold up to seven passengers.

# CHAPTER 6

**HEV COMPONENTS**

Components

This section summarizes the design specifications and component technology selections for the primary HEV configuration modeled in the study. The main technologies selected include the type of engine (e.g., spark ignition or diesel), electric motor 18 (AC induction, DC, or DC brushless permanent magnet), and batteries (nickel metal hydride or lead acid). Selection criteria included vehicle performance requirements, technology maturity, and prospective costs.

Component capabilities were then quantified as part of, and to support, the HEV performance modeling efforts. These efforts also identified key uncertainties in component specifications and evaluated the sensitivities of the results to these uncertainties.

HEV component specifications were quantified by applying the ADVISOR (ADvanced VehIcle SimulatOR) 2.2.1d HEV performance simulation model in an iterative fashion so that the predicted vehicle performance met targets, such as those shown in Table 3-3 for the mid-size HEVs. Some customization of the model was involved in these efforts. The ADVISOR program, driving cycles, and other details of HEV performance prediction calculations are documented in Section 3.3.1. Figure 3-1 shows a generic parallel HEV schematic to aid the understanding of the functional components and their specifications discussed below. Typically, as all-electric range increases (i.e. going from an HEV 0 to an HEV 60), the engine gets smaller and the motor and battery get larger. In addition, the HEV 20 and HEV 60 will have an on-board charger and cable to connect it to the power grid. An air conditioning compressor and a power steering pump are driven off the accessory drive.

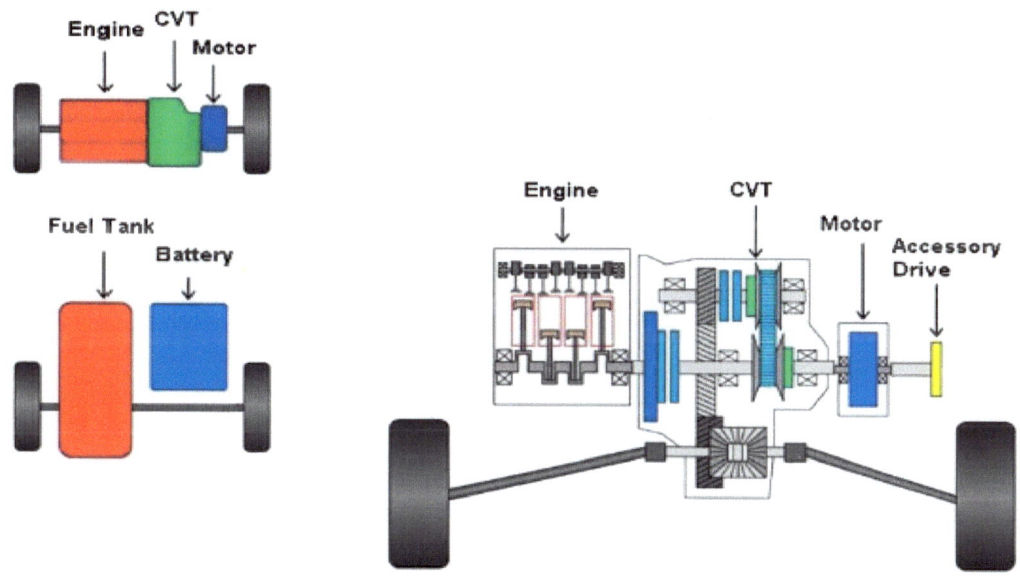

## 1. Gasoline Engine

The gasoline enginein aHEV is similar to that found in a conventionalICE vehicle. Gasolineengines inHEV s are usually much smalle rthan ones found in comparableconventional vehicles. Largerengines areprimarily heavier, requiring extraenergy during accelerations orclimbinginclinations; pistons along withothercomponents are heavier in a larger engine, which decreasethe efficiency andadd to the overall weight ofthe vehicle. The gasoline engine is the primary source of power for the vehicle, and the electric motor is the secondary source of power. The ToyotaPrius for example can operate in stand aloneelectricmodeatlow speeds(usuallyupto 15 mph), and can offer assistance during heavy

acceleration or when a power boost is required

## 2. Electric Motors

The electric motoris primarilyused to drive HEV s at low speeds, andassistthe gasoline engine

when additional power is required. The electric motor can even act as a generator and convert

energy from the engine or through regenerative braking into electricity, which is then stored in the battery. This functionality works as the electric motor applies a resistive force to the drivetrain which causes the wheels to slow down. The energy from thewheels then begin to turn theelectric motor, making it operate as a generator, converting this nonnallywastedenergythroughcoastingandbraking into electricity

## 3. Generator

In a series configuredHEV (discussed later) onlythe electric motoris connected to the wheels.

A series HEV has a separate generator which is coupled with the gasoline engine. The engine/

generator set supplies the electricityrequired bythe batteries, in tum feeding the electric motor. The coupled generator and engine maintain the efficient usage ofthe battery system during operation.

# CHAPTER 7

**PERFORMANCE OF HYBRID VEHICLES**

A central task of this study was to model the performance of a number of hybrid vehicle configurations, that is, combinations of common vehicle platforms and selected hybrid-electric designs. These models became the basis not only for predicting the vehicle fuel efficiencies and emissions discussed in this section but also for performing the cost estimation (Section 4) and customer preference analyses (Section 5). The configurations considered, and the main issues pertinent to their selection and interpretation, are discussed in Section 3.2. While the study examined several platforms, this report centers on mid-size cars. Other platforms (compacts and SUVs) will be part of a subsequent report.

When comparing different HEVs to each other as well as to conventional vehicles, the efficiency and environmental characteristics are of primary interest. Among the latter characteristics, those most relevant are the "well-to-wheels" emissions that include upstream (fuel-cycle) emissions in addition to vehicle tailpipe emissions. Upstream emissions are associated with the generation of the electricity used to charge plug-in HEV batteries and with the production, transportation, refining, and distribution of fossil fuels

In a hybrid vehicle the gasoline engine can be much smaller than that of a conventional car and Therefore is more efficient. Most cars need a relatively big engine to produce enough power for rapid acceleration of the vehicle. In a small engine, however, the efficiency can be improved by use of parts smaller and lighter by reducing the number of cylinders and the engine operating closer to itsmaximum load.

Below are some reasons for the efficiency of smaller engines compared to larger one:
• The larger engine is heavier, so that the car uses extra energy when accelerating or must face an incline;
• The pistons and other internal components are heavier, requiring more energy always up and down the cylinder;
• The displacement of the cylinders is greater, so that more fuel is required for each cylinder;
• Larger engines usually have more cylinders, each of which uses fuel when the engine is turned on, even if the car is not moving.

This explains why two cars of the same model but with different engines, may have different fuel consumption. If the two cars travelling on a highway at the same speed, one that has smaller engine uses less energy. Both engines need to produce the same level of power to propel the car, but the smaller engine uses less energy. The hybrid car uses a smaller engine, and when there is a need for increased efforts, the hybrid vehicle uses the electric motor.

Figure 1. Sales of HEV vehicles in the USA

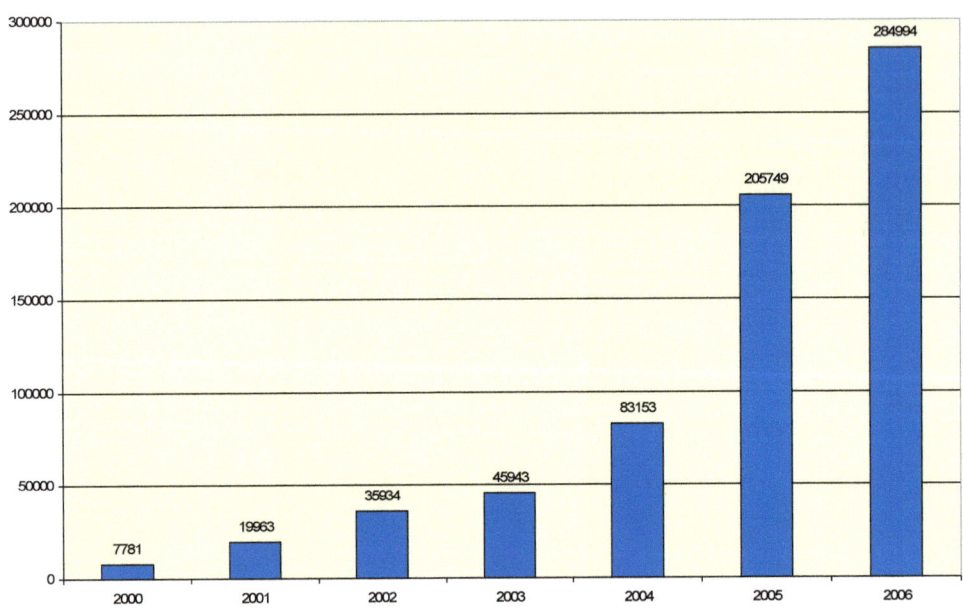

As far as regional breakdown is concerned for 2005, new hybrid vehicle registrations in California strongly outpaced all other states with 52,619 units, with Florida being second with only 10,470 units. Texas came in third with 9,632 vehicles; then New York came in fourth with 9,372 units; and Virginia rounds out the top five with 8,650 new hybrid vehicle registrations. Indeed Los Angeles remains the top metropolitan area for hybrid vehicles with 22,922 new hybrid vehicle registrations. San Francisco also kept its number two ranking with 15,828 units, followed by New York with 11,351 hybrids. Washington, D.C. came in fourth at 9,396 vehicles, followed by Boston with 3,641 new registrations.

For Japan, there are no official figures for sales of hybrid cars. The following data on hybrid vehicles in use have been published by JAMA, the OEM professional association:

Figure 2 : **Hybrid Vehicles in Use in Japan**

It is indeed very difficult to foresee market share for hybrid cars in the developed countries since it is an emerging technology competing with a well-established dominant design supported by a strong lobby. For the USA, there are many forecasts of market share published by professional experts as well as university sources. For a given year, they can vary considerably: from 11% (Mercer) to 30% (Polk) in 2015 for instance. Some are obviously over-optimistic, some rather pessimistic. Forecasts still vary significantly over time due to adjustments and change of assumptions within the relevant key variables (see 2.2).

Greene, Duleep & McManus (2004) carried out an impressive survey in the United States about perception on hybrid versus diesel powertrains in the light duty vehicle market and provided a synthesis of most professional market literature available. They elaborated scenarios for 2008, 2012 and beyond which show that market share for hybrids will remain limited and below usual expectations.

Figure 3 :**Market Share of Hybrid in 2012 in an "optimistic" scenario**

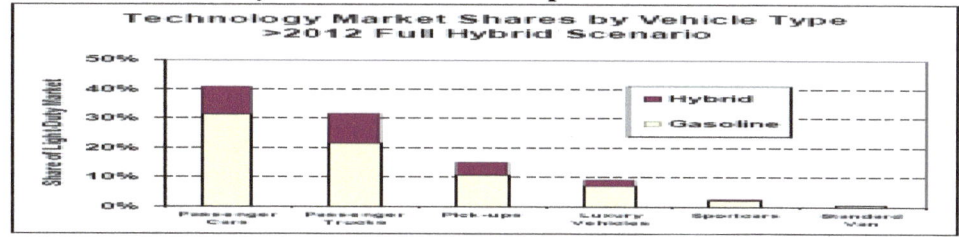

Forecast of HEV market share (US) by important consultant companies

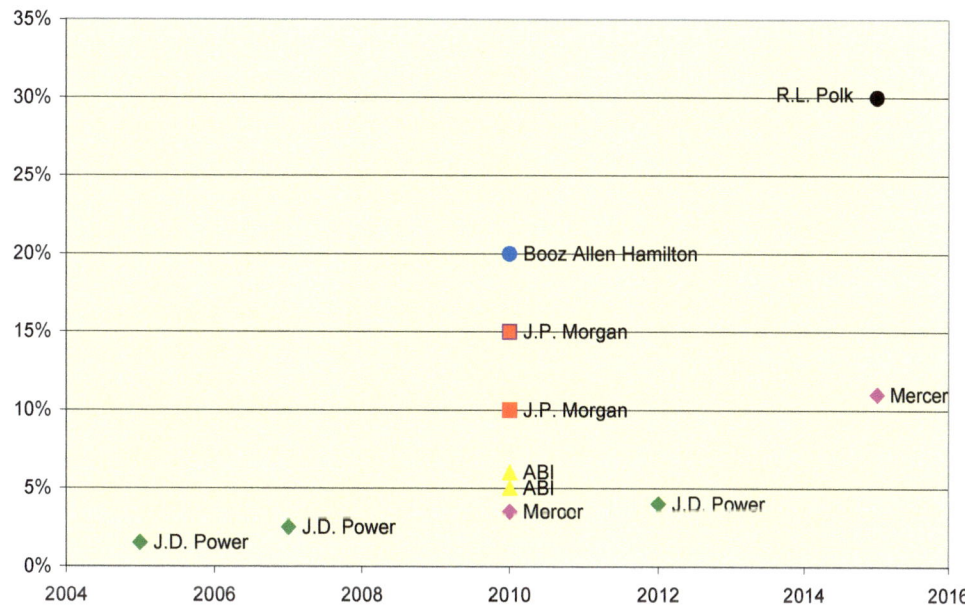

Teske's analysis of the HEV market in the US (2006) shows that future diffusion of HEV technology is driven by different variables. These can be grouped in three categories:

- the demand side of the market, including the perception of HEV technology among consumers, cost of ownership, maintenance constraints, and sociological reasons to choose HEV technology, such as the image impact of HEV technology;

- the supply side of the market, including the variety of HEV models offered, manufacturers' life cycle cost, characteristics of propulsion and fuel efficiency and emissions;

- as well as macro economic factors, which influence the market as a whole. Examples are the development of the fuel price, the availability and relevance of alternative technologies such as clean diesel and natural gas vehicles, regulation and taxes.

The most relevant include:

### Fuel Price

Being a major driver in the cost of ownership calculation, fuel price is considered the central variable in future market scenarios. However, strong HEV sales at times of high fuel prices in 2005 and 2006 also indicate a psychological dimension of this variable: Although those fuel prices did not necessarily make the purchase of a HEV worthwhile financially (Teske, 2006), the continuous increase of the fuel price itself pushed consumers to react.

### Growing environmental concerns

Environmental concerns become more and more a driver in consumers' purchase decisions. The broad public discussion on global warming due to $CO_2$ emissions, also by cars, is impacting buying decisions of consumers. In the USA, this is leveraged by the argument of energy independence. Driving a "green car" allows people to adopt a responsible and pro-active role in society. A new orientation towards ecology can be observed – in order to avoid remorse, the aspect of innovation gains importance over the classic interpretation of simply low fuel consumption.

### Fuel efficiency and emissions

$CO_2$ emissions are direct proportional to the fuel burned in the internal combustion process. Modern combustion technologies such as high precision gasoline injection allow reducing fuel consumption and lower harmful emissions such as CO, HC, and $NO_x$. HEV technology can help to lower emissions in different ways – by reducing a vehicle's fuel consumption through regenerative braking in stop-and-go traffic or by running the engine at highly efficient conditions.

### Improvement of energy storage

Energy storage is considered the key functionality of HEVs. Functionalities like regenerative braking or additional performance for acceleration not only require high capacities of energy – the electric energy needs to be stored and released quickly. Today's HEVs mainly use batteries based on Nickel-Metal-Hydride (NiMH) technology – which is relatively heavy and expensive. Ongoing research activities focus on the exploitation of Lithium-Ion (Li-Io) technology – a lighter and more powerful alternative. However, the technology was not yet transferred to cars due to its inability to handle the quick charging and release cycles of the cars. Capacitors, as seen in the BMW X3 Efficient Dynamics, cope with this requirement, but have limited capacities compared to currently used batteries.

*Regulatory and Tax Regime*

In some areas of the world, regulation clearly favours HEVs above conventional ICEs. This is especially the case in the eight US states who adopted the rules of the Californian Air Ressource Board (CARB): New York, New Jersey, Massachusetts, Maine, Connecticut, Rhode Island, Vermont, and California. The Zero-Emission-Vehicle (ZEV) Program aims at drastically reducing emissions related to air quality. Besides setting up future emission standards, governments influence car sales with short-term interventions. Several states in the US offer tax credits in order to support hybrid sales. The tax credit becomes effective upon purchase of an HEV. Tax incentives for HEVs are reported in at least 15 states of the US and three cities; another 13 states are considering these measures (www.usatoday.com, 2006). The tax credit amount is the sum of two factors: a fuel efficiency credit and a conservation credit. The fuel efficiency credit is based on a vehicle's increase in fuel economy over a 2002 comparable vehicle standard. The conservation credit is based on the estimated lifetime fuel savings. Thus, tax credits depend on the type of HEVs. In 2005, the highest tax credit of US$ 3,150 was granted upon purchase of a Toyota Prius.

Whereas the US tax systems clearly favour HEVs, European legislation is more divers: Currently, 11 EU member states have elements in their car and/or fuel taxation system that are totally or partly based on the car's $CO_2$ emissions and/or fuel consumption. With these tax systems being very different, they fail to send a clear market signals.

In order to lower the fragmentation of the EU market, the Automobile Manufacturers Association (ACEA) is calling for a harmonized, cross-Europe $CO_2$ tax on cars and alternative fuels. According to ACEA, $CO_2$ should be the key criterion for taxation to provide incentives to buy lower $CO_2$ emitting cars. Also, other than in some parts in the US, taxation should be technology-neutral to allow competition for the best solution.

The most common approach to compare fuel efficiency is to look at miles per gallon or at litres per 100 km. This approach can be also called tank-to-wheel approach (TTW), since it compares fuel efficiency after the fuel was produced and made available to the car. It is best practice when comparing cars using the same fuel with the same intrinsic energy.

Consumers buying decisions are more and more driven by a car's fuel economy. This is mainly due to increasing fuel prices. Consequently, if a HEV's higher initial purchase price can be compensated by savings on fuel, the investment is worthwhile financially. As Teske (2006) pointed out in an analysis on cost of ownership, other cost drivers are taken into account:

- Tax advantages, such as credits granted in the USA upon a HEV purchase, help compensating the higher purchase price.
- The expected resale price of a HEV is still difficult to predict due to the absence of experience. In 2006, most HEVs on the used-car market sold at relatively high prices, which can be explained by the high demand and the limited offer of new models during that time.
- To date, cost for maintenance and repair are very low, because most HEV manufacturers grant long term warranty periods on HEV components in order to avoid obvious

constraints of the new technology. E.g. Toyota's HEV related components are covered by an eight year warranty.

The analysis, conducted for the US market, showed that in general the purchase of an HEV does not allow significant financial benefits for consumers. However, an analysis of the situation in France by Capital (2006) shows a different conclusion.
Here, the hybrid option looks attractive, in particular thanks to the decrease in fuel cost and to the tax incentive (Table 1).

**Table 1 :** Total Cost over 4 Years

|  | Toyota Prius | | Honda Civic | | Lexus 4WD | | Citroën C3 | |
|---|---|---|---|---|---|---|---|---|
|  | Gasoline | Hybrid | Gasoline | Hybrid | Gasoline | Hybrid | Gasoline | Hybrid |
| Purchasing | 21,750 | 24,950 | 20,200 | 22,100 | 50,550 | 56,000 | 15,580 | 15,500 |
| Tax Incentive |  | -1,525 |  | -1,525 |  | -1,525 |  |  |
| Fuel | 8,730 | 4,501 | 8,140 | 5,085 | 12,165 | 6,544 | 7,298 | 5,849 |
| Insurance | 2,514 | 1,835 | 2,514 | 1,835 | 5,091 | 5,091 | 1,960 | 1,837 |
| **TOTAL** | **32,994** | **29,761** | **30,854** | **27,495** | **67,806** | **66,110** | **24,838** | **23,186** |

With financial benefits being not significant, there must be other drivers to the current success of HEVs in the US. A recent study conducted by Topline Strategy Group (2007) concludes that HEV drivers do not focus solely on financial performance. It is also some kind of environmental virtue driving the purchase decision. In the US, driving a hybrid car shows social responsibility to many consumers. This is also due to successful marketing of the hybrid label by the early adopters Toyota and Honda. It is not necessarily the expected lower $CO_2$ output of ones car; the image attached to driving a hybrid car also creates value for HEV drivers.

As highlighted above, the higher purchase price is a barrier to a faster diffusion of HEV technology. However, the price surplus differs significantly, taking into account regional and demographical preferences. As suggested by Teske (2006), the higher purchase price of an HEV can not only be compensated by savings over the car's lifetime; the technology also brings value of non-financial, i.e. qualitative nature. Examples are e.g. the surplus in torque delivered by the electric engines and the low noise during electric-only propulsion. Another important qualitative feature is the technology's image of social responsibility and sustainability. E.g. Toyota successfully shaped these characteristics by strong marketing activities in the US. These activities were leveraged by continuously rising fuel prices and an ongoing public discussion on climate change.

The issue of an "ideal pricing" for hybrid cars has been raised by Kishi & Satoh (2005). They suggest the following prices for ICE and hybrid cars on the Japanese market:

**Table 2 :** Prices for ICE and Hybrid Cars in Japan in US$

|  | Tokyo | | Sapporo | |
|---|---|---|---|---|
|  | ICE | Hybrid | ICE | Hybrid |

| | | | | |
|---|---|---|---|---|
| Minimum Price | 13,125 | 12,631 | 13,338 | 14,225 |
| Maximum Price | 28,535 | 28,007 | 26,677 | 25,424 |
| Standard Price | 21,222 | 20,932 | 20,148 | 19,978 |
| Reasonable Price | 15,716 | 15,154 | 16,049 | 17,498 |

Source: Kishi & Satoh, 2005. Exchange rate at 31/08/2006 is 1,000 yen = 8,523US$.

Some of the biggest OEMs promote HEV technology as the best alternative to increase vehicle efficiency. Most press articles and corporate literature published by HEV manufacturers and suppliers look at HEV with optimism. Opponents to HEV obviously attack the technology with strong arguments against its technical complexity and challengeable overall efficiency.

Indeed political pressures are also involved in the game. The three American OEMs, namely GM, Ford and Chrysler, recently urged President Bush to financially and politically support a national technological solution for hybrids, independent from the currently dominant solutions initiated by Toyota.

## Vehicle Performance Specifications and Efficiencies

This section explains the simulation approach, presents baseline mid-size vehicle results, and discusses additional vehicle configurations.

**Parallel HEV Component Technologies and Model Baseline Values**

| Vehicle Architecture | CV | HEV 0 | HEV 20 | HEV 60 |
|---|---|---|---|---|
| **Engine** | | | | |
| Base engine map | Lumina 3.1L | Prius Atkinson[a] | Prius Atkinson[a] | Prius Atkinson[a] |
| Base engine output (kW) Rated | 127 | 43 | 43 | 43 |
| Final engine output (kW) Rated | 127 | 67 | 61 | 38 |
| Scaled by | N/A | Max torque | Max torque | Max torque |
| **Electric Motor** | | | | |
| Base motor efficiency map | N/A | Precept 35 kW PM | Precept 35 kW PM | Precept 35 kW PM |
| Base motor peak output (kW) | | 35 | 35 | 35 |
| Final motor peak output (kW) [b] | | 44 | 51 | 75 |
| Continuous motor output [c] | | 19 | 22 | 32 |
| Scaled by | | Max torque | Max torque | Max torque |
| **High-Voltage Battery** | | | | |
| Base battery efficiency map | N/A | Ovonic HiPwr | Ovonic HEV-28 | Ovonic HEV-45 |
| System voltage (V) [d] | | 381 | 217 | 388 |
| Base specific power (W/kg)[e] | | 650 | 444 | 393 |
| Base specific energy (Wh/kg)[f] | | 37 | 49 | 71 |
| Final specific power (W/kg) | | 650 | 441 | 393 |
| Final specific energy (Wh/kg) | | 39 | 48 [g] | 71 |
| Final battery power (kW) | | 49 | 54 | 99 |
| Final battery energy (kWh) | | 2.9 | 5.9 | 17.9 |
| Peak specific power to energy ratio | | 16.9 | 9.1 | 5.5 |
| Scaled by | | Battery energy | Battery energy | Battery energy |

# CHAPTER 8

## *Design Methodology and Performance*

HEV and CV component and vehicle characteristics were modeled using the ADVISOR (ADvanced VehIcle SimulatOR) computer program developed by the National RenewableEnergy Laboratory (NREL) with support from Department of Energy (DOE). Each HEV wasconceptually designed by the WG as part of an iterative process to meet or exceed the performance of the baseline CV in several performance categories, including various acceleration, top speed, gradeability, minimum towing capability, and minimum range targets.

In addition, plug-in HEVs were asked to meet these performance targets with a battery discharged down to nearly 20% state of charge (SOC), the lowest SOC permitted in the interest of good battery cycle life. In a few cases, HEV performance parameters were relaxed somewhat if matching a specific CV parameter would have increased the cost of the HEV design greatly with only marginal useful gains for the vehicle owner/operator.

## Sustained Top Speed

The target for all HEVs was established at 90 mph, while a typical mid-size CV top speed was estimated to be approximately 120 mph. (The final HEV 0 designactually could sustain 120 mph, the HEV 20 could sustain 98 mph, the HEV 60 could sustain 97 mph and maintain 120 mph for about 2 minutes even with a low battery.)

## Gradeability

The HEV gradeability targets were 7.2% at 50 mph for 15 minutes and 7.2% at 30 mph for 30 minutes, while a typical mid-size CV gradeability is 7.2% at 50 mph for 30 minutes. (The HEV target is equivalent to climbing one of the longest and toughest grades in the world, the road to the top of Pike's Peak in Colorado, at 50 mph. All HEVs could also maintain freeway speeds on maximum Interstate Highway grades.)

## Passing Performance and Standing Acceleration

The target time to accelerate from 50 to 70 mph was increased from 4.8 seconds to 5.1 seconds while the 0-60 mph acceleration target was lowered from 11 seconds to 9.5 seconds, well within the 8 to 12 second range of 0-60 mph acceleration times of representative mid-size CVs. As shown in Table 2-1, HEVs exceed CV performance in 3 categories, with the CV slightly exceeding HEV performance in the 50-70 mph passing time category.

## Gasoline Range

All vehicles were designed to travel 350 miles using gasoline in the charge sustaining mode, but the HEVs' gasoline tank needed to have only about two-thirds of the CV's capacity.

**Acceleration Results for the Mid-Size Car**

| Vehicle Type | CV | HEV 0 | HEV 20 | HEV 60 |
|---|---|---|---|---|
| 0 to 30 mph, seconds | 3.5 | 3.1 | 3.0 | 3.0 |
| 0 to 60 mph, seconds | 9.3 | 8.7 | 8.9 | 8.9 |
| 40 to 60 mph, seconds | 4.6 | 4.2 | 4.3 | 4.3 |
| 50 to 70 mph, seconds | 4.5 | 5.2 | 5.2 | 5.2 |

**Trailer Towing**—All vehicles met the requirement to tow a 1,000 kg trailer.

### HEV Engine Stop/Starts

To minimize driveability issues, all HEVs were limited to 30 engine stop/starts on the FUDS3 driving cycle by adjusting their control strategy accordingly. (This strategy reduced fuel economy by about 10% relative to the maximum value that required 80 engine stop/ starts per drive cycle).

### High Speed Driving (HEV 20 and HEV 60)

At a low battery SOC, both plug-in HEVs exceed the modeling target to complete the federal test cycle for aggressive and higher speed (65-80 mph) driving (US061) twice in a row. All HEVs can do this indefinitely at low SOC, substantially exceeding the original expectations. In fact, there is enough battery capacity for the HEV 60 to complete this rather stringent test cycle for 40 miles using only the battery.

Even the HEV 20 was able to complete the 16-mile US06 cycle operating almost entirely (98%) in all electric (battery-only) mode.

### *Design Issues*

A number of design issues were identified in the efforts to model HEV performance. While there are probably solutions to each of them, they need further analysis. Section 4.6 includes a Preliminary discussion of the following and other design issues that should be examined for Successful commercialization of HEVs.

**Designs which only Include a Single Motor** (versus two motor designs). A single motor solution may produce unacceptable shift quality and unmanageable accessory drive and engine starting.

**Battery Pack Placement/Location** in the hybrid vehicles are a concern especially with larger battery packs such as in the HEV 60.

**Turtle" Light (Limited Battery Reserve Capacity)** could be a challenge for plug-in hybrids depending on which control strategy is selected.

**Cabin Heating.** It is assumed in this study that the engine would provide the heat for the cabin, but fuel economy would be sacrificed to perform cabin heating. Other forms of heat could be used such as Positive Thermal Coefficient (PTC) heating element at a cost.

**Battery Cooling.** Electric losses in the battery during charging and discharging generate heat that must be removed by flow of air or liquid coolant to keep the battery temperatures under control.

**Battery Life and Battery Replacement.** Under some driving scenarios, plug-in hybrid electric vehicles may require that the battery be replaced during the nominal lifetime of the vehicle. If the vehicle is driven over 100,000 miles or operated longer than 10 years, all hybrids might need a battery replacement. However, larger battery packs tend to accumulate more miles before replacement is needed.

__ **Control Strategies.** There are many control strategies than can be used with plug-in HEVs that can optimize fuel economy, emissions, and/or battery life. These should be examined in detail to determine the best possible strategy for plug-in HEVs.

# CHAPTER 9

## Vehicle Efficiency (Fuel Economy)

The hybridization of combustion engines with electrical energy storage devices into hybrid drive trains can reduce gasoline fuel consumption in two ways. All types of HEVs can make more efficient use of gasoline because hybridization permits not only the use of smaller engines operated more efficiently but also partial recovery of vehicle kinetic energy when the vehicle is decelerating or going down a hill. In addition, plug-in HEVs permit substitution of electricity as propulsion "fuel" for part of the gasoline.

Fuel economy can be defined in several ways, as shown in Figure 2-1. The gasoline only fuel economy applies to the CV and HEV 0 in all driving modes. It also applies for the HEV 20 and the HEV 60 whenever these plug-in hybrids are driven in the charge-sustaining mode, that is, with no net change in the energy content of the battery.

Electric-only fuel economy, expressed normally as miles per unit of electric energy is converted in this report to miles per (energy) equivalent gasoline gallon (mpeg) whenever a plug-in HEV is operating in electric-only mode; the energy equivalent calculations uses a conversion factor of 33.44 kWh per gallon of gasoline.

While plug-in hybrid electric vehicles can be operated for a given distance in electric-only mode (determined by battery capacity), trips since the battery was last charged that are longer than the HEV's all-electric range have mixed operation, i.e. some in electric-only mode and some in charge sustaining mode. The probability of a given HEV operating all its mileage in all-electric mode is referred to here as a mileage weighted probability (MWP).

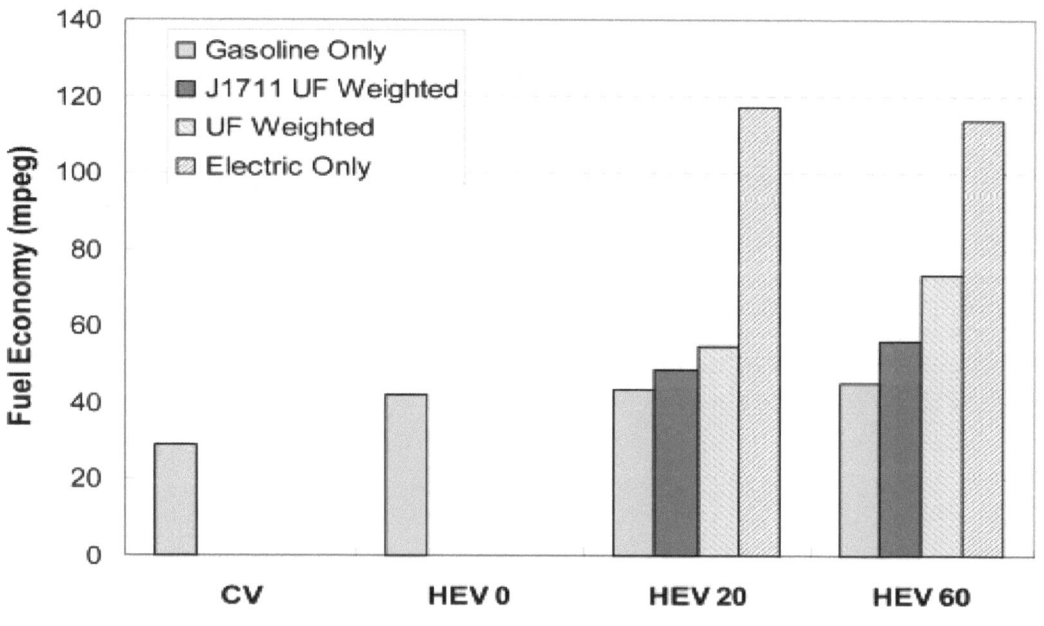

**Fuel Economy Comparisons for the Mid-Size Car**

35

Charging frequency also plays a part in determining the portion of annual miles that a plug-in HEV will operate in all-electric mode. The more often a plug-in hybrid is charged, the more Likely it is to travel a greater percentage of its annual miles in all-electric mode.

The SAE subcommittee also developed a recommended practice (J1711) that assumes the vehicle is just aslikely to start a trip with the battery fully charged as with the battery at a low SOC. This provides a case between charging every night and not charging at all. Other cases might include charging twice daily (at home and at work), or charging every other day. (See Section 3.3.1.4 for the effect charging frequency has on fuel economy.)

Gasoline includes methyl tertiary butyl ether (MTBE), an oxygenate made from natural gas

Electricity used to charge plug-in HEVs is generated by combined cycle natural gas fired power plants, in the assumption that charging will be at night with electricity produced at the margin (additional electricity produced on top of the current electricity needs most likely will be produced by natural gas fired power plants; therefore, electricity production is allocated to natural gas.

The energy used for fuel/energy production facility construction and vehicle construction is generally less than 15% of vehicle lifetime energy use.

1.Gasoline includes methyl tertiary butyl ether (MTBE), an oxygenate made from natural gas

2. Electricity used to charge plug-in HEVs is generated by combined cycle natural gas fired power plants, in the assumption that charging will be at night with electricity produced at the margin (additional electricity produced on top of the current electricity needs most likely will be produced by natural gas fired power plants; therefore, electricity production is allocated to natural gas.

3.The energy used for fuel/energy production facility construction and vehicle construction is generally less than 15% of vehicle lifetime energy use.

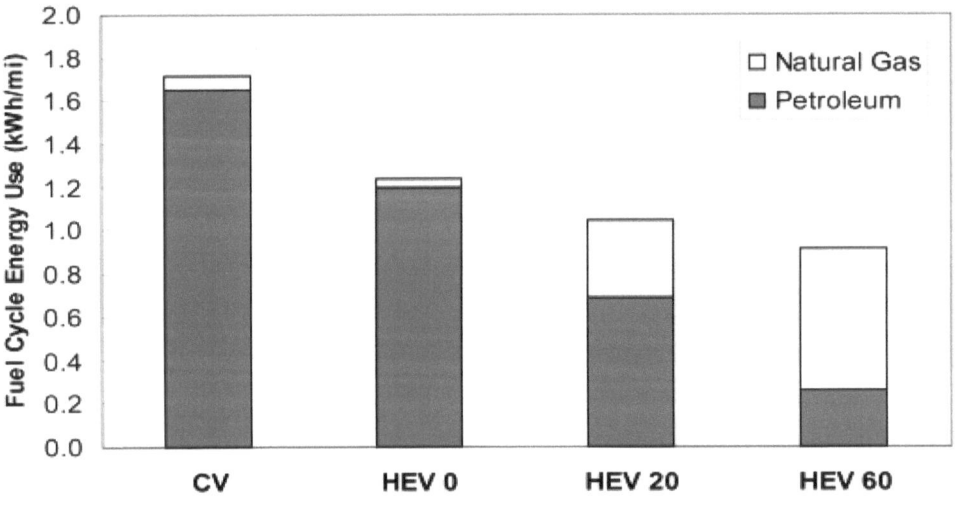

**Full Fuel-Cycle Energy Use for the Mid-Size Car for the Average Driving Cycle and**

# CHAPTER 10
**End-of-life Battery Test Methods**

The traction battery pack capacity of each HEV was characterized in accordance with test procedure ETA-HTP14, August 1, 2005,4 for static capacity and the HPPC test. To ensure consistency, all vehicle testing was performed in a temperature-controlled environment and with identical test protocol.

For static battery pack capacity testing, each end-of-life HEV demonstrated a reduced battery pack capacity (Figure 4). The two Civics demonstrated an average of 68.3% of battery pack capacity remaining, the two Insights an average of 84.6% of battery pack capacity remaining, and the two Priuses an average of 39.2% of battery pack capacity remaining.

The HPPC findings (Table 2) suggest a qualitative measure of the capability (range of working capacity) of each end-of-life HEV battery pack to meet a short-term, high-load demand that is representative of a typical drive cycle. Although in the future, an updated test procedure will require the HPPC test to be performed on new HEV battery packs, HPPC testing was not performed on the six subject HEVs (160,000 miles end-of-life fleet test) when the vehicles were new. The corresponding percent SOC at termination and the SOC step at which a discharge pulse was limited, as determined by HPPC testing, is displayed in Table 2. The lower the percentage SOC displayed in Table 2, the greater its capability is to meet the power demand of the motor controller at the discharge loads observed during the HPPC test. All batteries tested were capable of absorbing the charge pulses without reaching the voltage limit placed on them, including the charge pulse at 90% SOC. Therefore it is reasonable to conclude that the battery's ability to absorb energy had not degraded as a result of 160,000 miles of fleet testing. Also displayed in Table 2 are the pulse charge and pulse discharge currents used in the HPPC testing, as well as their corresponding C Rate (the current value divided by the battery's nominal C1 rating)

With the magnitude of the charge and discharge pulse established, the battery pack was subjected to a single pulse discharge and single pulse charge at each percent SOC level, starting at 90% and decrementing at 10% SOC intervals until the battery pack voltage reached an average of 0.8 volts per cell.

Between each test cycle, which consisted of one charge/discharge pulse at each percent SOC level, the battery pack was discharged at its C1 rate to reach the next 10% SOC interval. Upon reaching the termination criterion, the percent SOC and its equivalent Ah rating were recorded.

Although summarized above, a more detailed account of the HPPC test protocol included fully charging each battery pack by performing one C1 cycle as described in the static capacity test, followed by a one-hour rest period to allow for cell stabilization. At the end of this rest period, the battery was discharged at the C1 rate until it reached 90% SOC based on the nominal C1 Ah rating of the battery, (i.e. nominal C1 battery rating in Ah times 0.9). Immediately following a one-hour rest period, the battery pack was subjected to a ten-second discharge pulse of the discharge magnitude determined in the method above. A forty-second rest period was observed followed by a ten-second current charge pulse of the charge magnitude determined in the method above. After the discharge/charge pulse, the battery pack was discharged at a C1 rate until it reached 80% SOC, immediately followed by a one-hour rest period. This pulse sequence was continued, removing 10% of SOC each time, until the battery pack voltage averaged 0.8 volts per cell during the C1 discharge portion, at which point the test was terminated (the battery pack voltage was protection-limited to an average 0.8 volts per cell). If the voltage reached an average of 0.8 volts per cell at any point during a discharge pulse, the current was decreased until the pulse duration was complete. In this way, the battery pack was protected from over-discharge. An upper limit of 1.8 volts per cell was also placed on the battery during the charge pulse portion of the test to protect the battery from overcharge.

## End-of-life Battery Test Results

Since the initial traction battery pack capacity of each HEV was not determined when the vehicle was new, the characterization results obtained from the end-of-life testing were compared to the nominal (manufacturer) rated battery capacity.

For static battery pack capacity testing, each end-of-life HEV demonstrated a reduced battery pack capacity (Figure 4). The two Civics demonstrated an average of 68.3% of battery pack capacity remaining, the two Insights an average of 84.6% of battery pack capacity remaining, and the two Priuses an average of 39.2% of battery pack capacity remaining.

The HPPC findings (Table 2) suggest a qualitative measure of the capability (range of working capacity) of each end-of-life HEV battery pack to meet a short-term, high-load demand that is representative of a typical drive cycle. Although in the future, an updated test procedure will require the HPPC test to be performed on new HEV battery packs, HPPC testing was not performed on the six subject HEVs (160,000 miles end-of-life fleet test) when the vehicles were new. The corresponding percent SOC at termination and the SOC step at which a discharge pulse was limited, as determined by HPPC testing, is displayed in Table 2. The lower the percentage SOC displayed in Table 2, the greater its capability is to meet the power demand of the motor controller at the discharge loads observed during the HPPC test. All batteries tested were capable of absorbing the charge pulses without reaching the voltage limit placed on them, including the charge pulse at 90% SOC. Therefore it is reasonable to conclude that the battery's ability to absorb energy had not degraded as a result of 160,000 miles of fleet testing. Also displayed in Table 2 are the pulse charge and pulse discharge currents used in the HPPC testing, as well as their corresponding C Rate (the current value divided by the battery's nominal C1 rating).

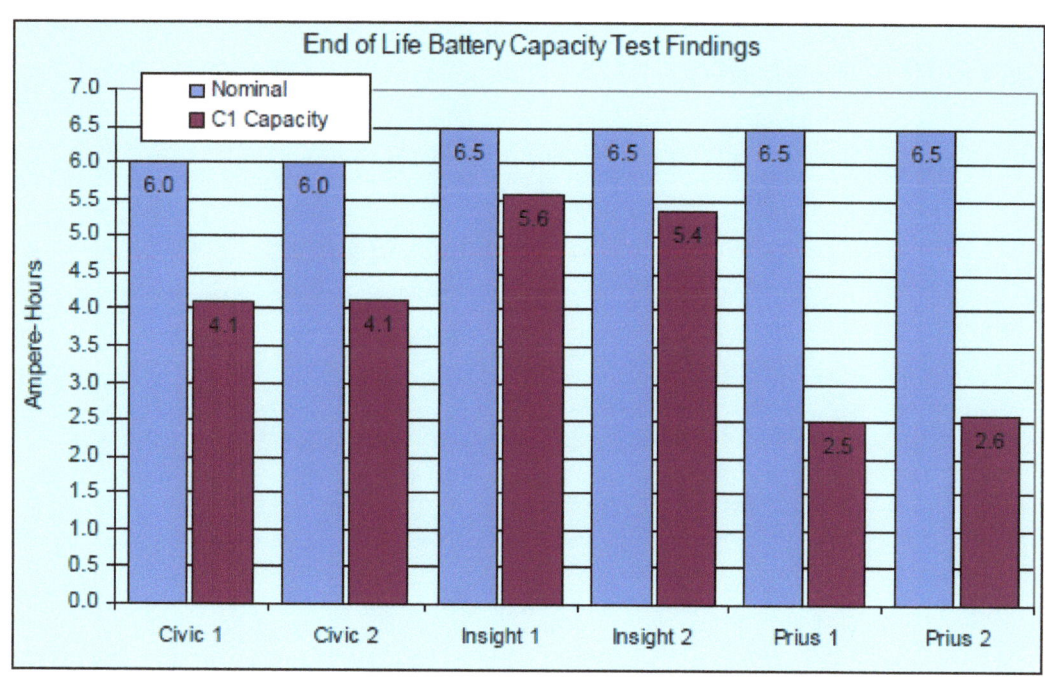

Figure 4. End-of-Life HEV battery capacity test findings

Table 2. End-of-Life HPPC test findings of percent SOC at test termination.

| End-of-Life HEV | Percent SOC at Discharge Pulse Voltage Limit | Percent SOC at test end | Charge Pulse Amps / C Rate | Discharge Pulse Amps / C Rate |
|---|---|---|---|---|
| Civic 1 | 30% | 23.0% | 71.6 / 11.9 | -68.4 / -11.4 |
| Civic 2 | 50% | 41.6% | 71.6 / 11.9 | -68.4 / -11.4 |
| Insight 1 | 10% | 5.9% | 78.7 / 12.1 | -65.7 / -10.1 |
| Insight 2 | 10% | 8.2% | 78.7 / 12.1 | -65.7 / -10.1 |
| Prius 1 | 60% | 52.4% | 53.4 / 8.2 | -78.9 / -12.1 |
| Prius 2 | * | 51.2% | 53.4 / 8.2 | -78.9 / -12.1 |

## DISCUSSION OF END-OF-LIFE TESTING RESULTS

Based on comparing the limited end-of-life test results to the baseline performance (new vehicle) test results, the data suggests that powertrain reliance on obtaining propulsion energy from the battery pack increases in the Insight and Civic (Figure 3) as they age. Coupled with the fact that all of the end-of-life HEVs tested had a decrease in battery pack capacity (Figure 4), these factors are not the only variables influencing changes in fuel efficiency, as the Prius demonstrates by having a 60.8% reduction (highest change of the three Phase II vehicles) in battery pack capacity and an average decrease in fuel efficiency (Phase II, AC-off and AC-on combined) of only 1.3% (lowest change of the three Phase II vehicles).

The Insight data also suggests that battery capacity reduction and a greater reliance on the battery pack to supply propulsion energy does not play a dominant role in affecting fuel efficiency as shown by the end-of-life Insight's 15.4% reduction (least change of the three, Phase II vehicles) in battery pack

# CHAPTER 11

## CONCLUSION

However, HEV technology is only one subset in the wide range of measures to lower fuel consumption. The main aims of current vehicle development, to increase fuel efficiency while lowering weight and costs can be solved alternatively as shown with current diesel models or intelligent cost-efficient solutions as the BMW example shows.

Consumers' buying decisions are driven by factors different than financial ones only. These are dynamic in nature due to an ongoing public discussion on climate change and related changes within macro-economic factors such as taxes. Anticipating these changes will be central to successfully marketing cars in the future.

This article comes to the conclusion that HEV will probably gain a significant market share in Japan and the United States due to market pressures, sustained by political lobbying, but will remain limited elsewhere in the mid-term. One of the clearer uncertainties is the potential effect of a development of HEV market in China where environmental issues are of significant concern. Chinese OEMs could see the HEV as a solution for their growth in a very competitive globalized market. A positive driving force to HEV is certainly the difficulties identified by fundamental and applied research and development units with OEMs or component suppliers as well as public laboratories in developing more sophisticated technologies such as fuel cells or batteries for full electric drive trains. Such alternatives require breakthrough disruptive innovations and would probably emerge in the very long future, i.e. at least not before ten or even twenty years. In the meantime, HEV technology might have a real future.

### Limitations

This article has reported the results of an investigation into the latent dimensions used to evaluate car purchase by consumers who were considering buying a hybrid-electric car. These dimensions were compared to a group of respondents who were thinking of buying a conventionally fuelled vehicle. These two groups of car buyers were actively evaluating automobiles for purchase in the next twelve months and had saved money or had access to funds to purchase a vehicle.

We found that respondents choosing hybrid vehicles evaluated the purchase differently from buyers choosing conventional vehicles. At least in this sample, hybrid buyers were mainly concerned with whether the car would improve their social standing and personal image. This finding is consistent with that reported by Griskevicius, Tybur & Van den Bergh (2010). Buyers of conventional cars were more concerned with the car's functionality, cost and quality and were less concerned wither the car made them socially popular. What this article added to the literature is that we now have an idea of the relative importance of the evaluative criteria for petrol-electric hybrid vehicle purchase. Our model also provides a map of how the items have loaded into these evaluative dimensions.

Although the sample consisted of many university graduates, the sampling was done in the workplace. It just happened that many of the researchers' colleagues were graduates. This population was young (22-30) and were the prime market for automobile manufacturers.

It appears that hybrid cars, at least for our sample, appear to be purchased for social and reasons and not by people who genuinely care for the environment. This finding is reasonable as other research (e.g. Heffner, Kurani & Turrentine 2007) have reported that the most environmentally sensitive consumers preferred to abstain from driving. Our sample of hybrid car buyers is highly influenced by their reference groups. These groups seem to dictate the consumption behaviour of Prius buyers.

This strong influence of groups can be utilized by social marketers to shift social behaviour increase adoption of more environmentally friendly cars. In this case, a suitable model appears to be the diffusion of innovations model. Although rarely used by social marketers (Lefebvre 2000), this model can be used to promote the orderly adoption of hybrid vehicles. The way hybrid automobiles are bought, driven by social influence, provides positive answers to four areas necessary for successful diffusion of an innovation (Oldenburg, Hardcastle & Kok 1997). These areas are: does it fit into the audiences' lifestyle and self-image? Is the new behaviour better than current behaviour? Can it be trialled before commitment? Can it be easily and clearly understood? Finally, can the behaviour be adopted with minimal risk?

This externally motivated group (perhaps early adopters) can be influenced to achieve a critical mass of adoption for low-emissions vehicles. Social marketers can do this by using reference group appeals. In this case, it is important to achieve a salient positioning for the concept of low-emissions vehicles so that it appeals to more than one agency, and to later adopter groups that may be less prone to social influence (see results for buyers of conventional cars). It is expected that in order to achieve widespread adoption, the agencies that must be influenced include the customer, policy makers (e.g. government and regulators) and the community at large. It is only with this acceptance that the utilisation of low emissions vehicles will reach a critical point that it provides a positive effect on the environment.

www.ingramcontent.com/pod-product-compliance
Lightning Source LLC
Chambersburg PA
CBHW050839180526
45159CB00004B/1960